ISBN 978-1-332-21066-4
PIBN 10298733

This book is a reproduction of an important historical work. Forgotten Books uses
state-of-the-art technology to digitally reconstruct the work, preserving the original format
whilst repairing imperfections present in the aged copy. In rare cases, an imperfection in
the original, such as a blemish or missing page, may be replicated in our edition. We do,
however, repair the vast majority of imperfections successfully; any imperfections that
remain are intentionally left to preserve the state of such historical works.

THE
WATCHMAKER'S AND JEWELER'S
HAND-BOOK:

A CONCISE YET COMPREHENSIVE TREATISE

ON THE

"SECRETS OF THE TRADE."

A WORK OF RARE PRACTICAL VALUE TO

WATCHMAKERS, JEWELERS, SILVERSMITHS,

GOLD AND SILVER-PLATERS, Etc.

BEING DESIGNED AS A RELIABLE BOOK OF REFERENCE, IN WHICH MAY BE FOUND
IN INTELLIGIBLE FORM, THE BEST AND MOST APPROVED PROCESSES KNOWN
IN THESE IMPORTANT TRADES, TOGETHER WITH OTHER VALUABLE
MATTER, ADAPTING THE WORK TO THE REQUIREMENTS
OF A GENERAL WATCH, JEWELRY, AND
FANCY GOODS BUSINESS.

By C. HOPKINS.

————o○§◦o○§◦o○————

LOUISVILLE, KY.:
PUBLISHED FOR THE AUTHOR.
PRINTED BY JOHN P. MORTON & CO.
1866.

PREFACE.

THERE are few trades more difficult in which to acquire accuracy and skill than watch repairing, and perhaps few in which greater diversity in the modes of operating exist. This arises from the extreme delicacy of the instrument to be repaired, the numerous and often for a time imperceptible contingencies that arise to mar or impede its operation, and the absence of sources of correct knowledge as to the best modes of overcoming these diversified contingencies. Various works have been published on the subject of Chronometry, in some of which, by the aid of diagrams, the peculiar structure of the different instruments employed in noting time, as Chronometer, Duplex, Lever, Horizontal and Cylinder Escapement Watches, have been accurately and scientifically explained. But while these works have been found of great value to manufacturers of watches, in the simple daily requirements of watch repairing they have been of little practical avail. And, so far as the author is aware, no work, aiming to meet the special requirements of this important trade, has been attempted until the present time.

It is a well-known fact that all equally good workmen, in the main, and equally acquainted with the general structure of the watch, are not equally with each other skilled or successful in performing the same thing or things. And, other things being equal, this will invariably be found the result of disparity of knowledge of what are usually denominated the "secrets of the trade," or the best modes of operating on particular or specific points. The practical part of watch repairing, as of other important trades, can only be learned at the bench. But while the sources of knowledge relating thereto are restricted to verbal or oral instruction, it follows, as a necessary consequence, that imperfections and great disparity of knowledge in these regards must prevail, and that new or improved processes, the results of individual skill and ingenuity, must, from the very nature of things, remain from year to year, and perchance from age to age, in exclusive possession of the few, while the many, for want of available sources of information, are excluded from their benefits altogether.

No better proof of this is needed than the well-known fact that some of the processes known and practiced in Europe for a long series of years past are still unknown in this country, or known only to here and there an individual, who from fortuitous circumstances has chanced to be more highly favored than the great mass of his cotemporaries in the trade. And the same observation applies with equal force in regard to processes in this trade—the results of American as of European skill and ingenuity.

Therefore, in prepairing the Watchmaker's and Jeweler's Hand-book for publication, the author, while avoiding every thing of a merely historic or descriptive character, has aimed at supplying to all, as far as possible, the means of acquiring a correct knowledge of the various specialties above referred to; which a long practical experience, under peculiarly favorable circumstances and extensive travel, with a special view of collecting information on these subjects, enable him to do to a much fuller extent than under other circumstances would have been at all possible. The value placed upon some of these processes by others may be learned from the fact that, in more instances than one, gentlemen who understood and were practicing single ones of the many here described, have said to the author that even $100 would be no temptation to them to part therewith, and thus be forced to return to their previous modes of operating.

With these are given a large number of other processes and formulas, which, it is believed, will be found of no less practical value in the departments to which they relate than will those relating to watch repairing, in that department.

The process for gold and silver-plating, without a battery, it is believed is one of the best, if not the very best known for general use; while the various steps to be taken therein are so carefully defined that, by following the directions, any one, however inexperienced, may operate therewith with entire success. Similar remarks, with equal justice, might be made with respect to many of the other formulas. It is not, however, presumed nor supposed that all the processes herein described will be found new to all into whose hands this book may chance to fall; but believing there are few, if any, who may not find therein sufficient of practical value to warrant its purchase by them, the author confidently, yet very respectfully, commends his work to the favorable notice and patronage of those for whom it is specially designed.

WATCHMAKER'S AND JEWELER'S
HAND-BOOK.

PART FIRST.

EMBRACING PROCESSES WHEREIN MOST OF THE ARTICLES USED ARE
EMPLOYED IN LIQUID OR SEMI-LIQUID FORM.

Section I.—Improved Processes of Gold and Silver Plating, without a Battery.

PREPARATORY SOLUTIONS.

SOLUTION No. 1.—RECIPE—Take enough *rain-water* to cover, by immersion, whatever article or articles you may desire to plate, and dissolve in it sufficient *common potash* or *concentrated lye* to make the solution strong enough to bear up an egg or small potato. Let it settle until quite clear, and then pour off into a suitable glass or earthen vessel for use.

SOLUTION No. 2.—RECIPE—Nitric acid, $\frac{1}{2}$ oz.; muriatic acid, $\frac{1}{3}$ oz.; sheet zinc, $\frac{1}{4}$ oz., or as much as will dissolve; muriate of potash (fused), $\frac{1}{7}$ oz.; sulphate of iron, $\frac{1}{4}$ oz.; sulphuric acid, $\frac{1}{9}$ oz.; sulphuric ether, $\frac{1}{5}$ oz.

Put the nitric and muriatic acids together into an open glass or china vessel, and dissolve in them the zinc, muriate of potash, and sulphate of iron; then add the sulphuric acid and ether, and let it stand undisturbed in the open air for twenty-four hours. During this time a clear yellow liquid will be formed, which drain off carefully, and bottle for use.

GOLD SOLUTION.

FOR SIXTEEN-CARAT PLATE.

RECIPE—Nitric acid, 1 oz., muriatic acid, 2 oz., (aqua regia); silver (pure), $\frac{1}{4}$ dwt.; copper (pure), $\frac{1}{4}$ dwt.; gold coin, five dollars (about 5 dwt.)

Dissolve the gold, silver, and copper together in the aqua regia, and when solution is effected—which, if the acids are good, will generally be accomplished in from half an hour to one hour, (the silver in this case will not be taken up into solution with the gold and copper, but will remain at the bottom of the vessel in the form of a white powder—muriate or chloride of silver)—prepare the following, and add it by degrees to the solution:

RECIPE—Sulphate of iron, 1 oz.; borax, $\frac{1}{2}$ oz.; fine table-salt, 1 dwt.; pure rain-water, 1 qt.

In order to dissolve these articles readily, use *hot* water, and powder the borax and sulphate of iron. The office of this solution is to precipitate the gold and copper. Let it now stand about six hours, or until fully settled; then pour off carefully, and refill the vessel with hot rain-water. Let it again settle, and drain as before, and continue thus to wash and drain until no acid taste nor smell can be perceived; then drain fully, and add 1$\frac{1}{8}$ *ounce cyanuret potassa* and 1 *quart of hot rain-water*. In twenty-four hours it will be ready for use. The substitution of a little platinum as an alloy for the gold, in place of silver and copper, as above, will be found to operate very satisfactorily, and to make a much harder and more durable gilding or plate than can be produced with gold only, or with gold, silver, and copper combined.

To make *pure gold solution*, simply omit or leave out the silver and copper from the foregoing formula; or a lighter colored, pure gilding, as for watch movements, etc., if preferred, may be made by precipitating the gold in the proportions above given, with 1$\frac{1}{2}$ oz. sulphuret potassa dissolved in 1 qt. hot rain-water, in place of the sulphate of iron, borax, and table-salt.

A darker colored plate, when desired, may be produced by adding to the gold solution a little dragon's blood and iodid of iron in the proportion named in sec. 3. And a lighter shade, by adding from time to time a few drops of silver or platinum solution. The dragon's-blood and iodid of iron may be first dissolved in a little alcohol, and then added to the solution as required.

SILVER SOLUTION.

RECIPE—Best nitric acid, 2 oz.; distilled or pure rain-water, $\frac{1}{4}$ to 2 oz.; silver, 1 oz.

First roll or beat the silver pretty thin, cut into strips and drop it into the acid; then add the smallest quantity of water named in recipe, and let stand for a few minutes. If solution does not readily commence, add a little more water, and continue thus to add water, a little at a time, until the proper point is reached. From twenty minutes to one hour is usually the time required to dissolve silver by this process, irrespective of quantity. If from any cause, after dissolving for a time, the action of the acid on the silver should cease, the addition of a little more water will set it to work again usually with increased rapidity. When solution is fully effected, fill the vessel with rain-water, and add a large table-spoonful of fine table-salt; shake or stir it up well, and let it settle same as in the case of gold. When settled, pour off carefully, and refill the vessel with rain-water; let settle and drain as before, and continue thus to wash and drain until no acid taste can be perceived; then drain fully, and add 1½ oz. cyanuret potassa and 1 qt. clean rain-water. In twenty-four hours it will be ready for use. The nitro-muriate of silver, now undissolved at the bottom of the vessel, will serve to feed the solution, or restore it again to full strength after it has been used. The same observation applies in this regard as well to the gold as silver solution. A stronger solution may be made at pleasure, and one that will plate more rapidly by using a larger quantity of cyanuret potassa. But this is not desirable, and should be as a general rule avoided; as a better and firmer plate is always produced by using barely sufficient cyanuret potassa to answer the desired purpose.

GENERAL DIRECTIONS FOR USE.

First. Immerse the article to be plated, whatever it may be, for about fifteen miutes in Preparatory Solution No. 1. Then clean it off thoroughly with Spanish whiting and rinse in clean rain-water, and while still wet attach to it a piece of sheet zinc, and immerse it in the plating solution. After it has been in this solution for about one minute take it out, and again clean off with whiting as at the first; then return it to the solution; and when a heavy plate is desired, clean off in this way two, three, or more times during the process of plating.

Second. When the article to be plated is composed of iron, steel, lead, pewter, or block-tin, in whole or in part, after treating it as per directions for Solution No. 1, and before putting it into the plating solution, apply Solution No. 2 with a camel's-hair brush or with a bit of sponge to the entire surface, or to such part or parts only as you desire the plate to adhere to. The zinc should always come into direct and positive contact with the article being plated, but may be attached in any way most convenient to the operator. Probably the best mode, however, is to form a hook or loop on a strip of zinc, and with this suspend the article in the plating solution.   To prevent stain or discoloration from its contact with the article, the position of the zinc should be occasionally changed or shifted.   The thickness of the plate will of course depend upon the length of time the article remains in the plating solution.   This may be varied from a few minutes to a whole day, or longer if desired.   The solutions should be used at a temperature of between 70° and 80°, but in warm weather, or in a warm room, they will approach sufficiently near the temperature for the purpose.   In all processes of plating, either with or without a battery, it is a matter of the very first importance, and one that should never be carelessly or imperfectly attended to, to have the foundation properly prepared, or in other words, to have the article to be plated entirely and absolutely free from all grease, wax, dirt, stains, discolorations, etc., of whatever character.   In most cases, where other means fail, a thorough application of the gold or silver powders (see sec. 4) before putting the article into the plating solution, will lay the foundation for a firm, reliable plate.   The very best chemicals should always be used in preparing the various solutions, and care should be taken not to allow them to become mixed with any kind of foreign matter.   *For polishing*, fine jeweler's rouge, or Spanish whiting from which the gritty matter has been removed, will answer a good purpose; but a better finish with less loss of plate is always produced by properly burnishing.

To burnish gold, use, when practicable, a blood-stone or agate burnisher; and for silver, a finely polished steel burnisher, dipped, from time to time, while using, in a strong solution of soap and rain-water.

Section 2.—To Plate with a Battery.

Make the solutions the same as for plating *without* a *battery*. Then, after the article has been properly prepared, as per sec. 1, attach it to the positive pole of the battery, and with a piece of silver or gold, as the case may be, attached to the negative pole, immerse both together in the solution.   The solution, under the influence of the electric current, will dissolve the gold or silver attached to the negative pole of the battery as fast as it is deposited on the article being plated.   The two poles of the battery, with their attachments, should be placed opposite to each other in the solution, but not be brought into direct contact.   In all other respects, the same rules should be observed in this as in sec. 1.

Section 3.—To make Reddish-colored Gilding, of Twelve, Fourteen, Sixteen, or Eighteen-carat fine.

RECIPE—2 qt. rain-water; 1 oz. dragon's-blood; 1 oz. cyanuret potassa; 5 gr. iodid of iron. Put the dragon's-blood and water together into a suitable vessel, and boil down to one quart; then strain it, and add the cyanuret potassa and iodid of iron.

DIRECTIONS FOR USE.—Cleanse the article properly, and attach it to the positive pole of your battery, and to the negative pole attach a piece each of pure gold and copper, and immerse both together in the solution, as per sec. 2. The proportions of gold and copper used will determine, to some extent, both the quality and color of the gilding. A beautiful color may be given by this process, but is not adapted for heavy plating. It operates more satisfactorily, however, if the solution is used warm than when cold.

Section 4.—To make Gold and Silver Powders, for light Gold and Silver plating—by rubbing process—on Brass, Copper, Oreid, German Silver, etc., and for renovating Gold and Silver Plate of all kinds.

RECIPE—Dissolve gold and silver in the same manner and in the same proportions as in sec. 1, and after washing out the acids as there directed, to 1 oz. of silver or 5 dwt. of gold, add 3 oz. cyanuret potassa and 1 pt. of clean rain-water. Let stand until the chlorid of gold or silver, as the case may be, is all dissolved; then mix the solution with 1 lb. clean Spanish whiting, and evaporate to dryness in the open air. When dry, reduce to a fine powder, and bottle for use. In making the gold powder, add a little yellow ocher, to give color, and to distinguish it from the silver.

DIRECTIONS FOR USE.—For plating, dampen the powder with rain-water to the thickness of paste, and apply it with the finger, a rag, or a bit of cork; and for renovating, wet in like manner to about the consistency of cream, and apply with wash leather or a soft brush. On frosted work, either gold or silver, *pat* with the brush, instead of rubbing in the ordinary way.

After applying the powders as above, rinse clean in rain-water, and dry with a soft linen cloth or in fine sawdust.

Section 5.—To Frost and Gild Watch Movements, etc.—Swiss Process by first coating with Silver.

1. HOW TO PREPARE THE SILVER.—RECIPE—$1\frac{1}{2}$ dr. nitric acid; 2 dwt. pure silver; a little distilled or clean rain-water.

When solution of the silver is effected, add about 2 oz. clear rain-water, and immerse in it a piece of clean sheet copper. The silver will collect upon the copper, but may be readily scraped off. (See sec. 5, part 3.) When the silver in solution has been thus all precipitated to the bottom, collect it by straining through filtering paper, and dry without washing in the open air. When dry, scrape it carefully together from the paper, and bottle for use.

2. HOW TO BE APPLIED.—Stone off properly the article to be frosted, and pin it to a piece of cork, to prevent chafing; then mix together 1 part silver, prepared as above, with 6 parts fine table-salt, and sufficient rain-water to form the whole into a thin paste, which apply to the article with a watch or plate-brush, giving to the brush a circular motion while applying. When in this way a sufficient quantity of silver has been made to adhere to the article, place it, still on the cork, in a shallow vessel, and barely cover it with strong or sour beer; and while in this position, take a fine scratch-brush (a brush made by binding together a bundle of very fine brass or steel wire) and scour it thoroughly, giving the same circular motion to this as to the plate-brush first used. Then wash off and rinse clean in alcohol, and it is ready for gilding.

3. DIRECTIONS FOR GILDING.—After frosting as above, attach to the article a strip of sheet zinc, and immerse it for a short time in the gold solution, (see sec. 1,) and then clean off with a soft brush and cream tartar and rain-water, and dry in fine sawdust.

NOTE.—A better color is generally produced by light rather than heavy gilding, and a solution alloyed with a very little platinum, added to the gold while in course of preparation, will greatly increase the durability of the gilding.

Various other modes or processes of frosting are employed by different parties, among which may be named the following:

First. Immerse or sink the article by means of a bit of peg-wood for a short time in a mixture of nitric acid and fine table-salt, or about equal proportions of nitric and sulphuric acids and fine table-salt.

Second. Sprinkle over the article a thick coating of pulverized charcoal, and then dampen it with nitric acid, or with nitric and sulphuric acids combined.

Third (and simplest mode). Dip the article quickly several times in succession into the chain-dip solution, (see sec. 6,) holding it out each time barely long enough to allow the air to act upon it; then wash and rinse thoroughly, and dry in sawdust.

In each of the foregoing modes of frosting, after removing from the acid, the article may be placed in strong beer, scoured with a scratch-brush, and gilded with gold, same as in the latter part of the Swiss process.

Section 6.—Chain-Dip Solution, for Brass Chains, etc.

Removes instantaneously all stains or discolorations, and gives to the article a perfectly *bright* and *new* appearance.

RECIPE—$2\frac{1}{4}$ oz. sulphuric acid; 2 oz. nitric acid; 2 oz. rain-water; 1 dr. saltpeter.

Mix together in a glass bottle, and let stand for a few hours. Apply by dipping the article into the solution quickly, and then at once wash off thoroughly, and rinse in clean rain-water and dry in sawdust; or, if convenient, after washing in water, rinse in alcohol, or in a weak dilution of spirits of ammonia, and rain-water.

Section 7.—To make Etruscan Gold Coloring, and to renew Tarnished Etruscan Jewelry.

RECIPE—1 oz. alum; 1 oz. fine table-salt; 2 oz. saltpeter, (powdered); hot rain-water, sufficient to make the solution, when all is dissolved, about the consistency of thick ale; then add sufficient muriatic acid to produce the color desired. No absolute rule can be given for the proportion of acid to be used in this process, as it requires in different cases to be varied according to circumstances. Hence, the degree of success must always depend, to a greater or less extent, upon the judgment and skill of the operator. The article to be colored should be from fourteen to eighteen carats fine, and composed of pure gold and copper only, without admixture of other alloy, and be free from coatings of tin or silver solder. The solution acts more promptly and satisfactorily when used warm than cold; and, in order to produce the best possible color, should be made fresh each time of using; but any desired number of articles may be colored at the same time.

The principle on which this solution acts is to eat out the copper alloy from the surface of the article, leaving thereon pure, frosted gold only.

After coloring, wash off first in rain-water, then in alcohol, and dry, without rubbing, in fine, clean sawdust. Fine Etruscan jewelry, that has been defaced or tarnihsed by use, may be perfectly renewed by the same process.

Section 8.—Pickle, for Frosting and Whitening Silver Goods.

RECIPE—1 dr. sulphuric acid; 4 oz. water. Heat the pickle, and immerse the silver in it until frosted as desired; then wash off clean, and dry with a soft linen cloth, or in fine, clean sawdust. For whitening only, a smaller proportion of acid may be used.

Section 9.—To remove Discolorations from Gold Rings, etc., tarnished by Heat.

Use a pickle of water and sulphuric acid in the same way, but somewhat stronger, than for frosting silver; or for cheap gold use a strong solution of cyanuret potassa and rain-water.

Section 10.—To remove Tarnish from Electro-plated Goods, without scratching or injury to the Polish.

RECIPE—2 gal. rain-water; ½ lb. cyanuret potassa.

Dissolve and put into a suitable covered vessel for use, or, what is better, into a stone jug or jar, and closely cork.

DIRECTIONS FOR USE.—Immerse the tarnished article in the solution from one to ten or fifteen minutes, or until the tarnish has been removed, and no longer; then wash and rinse it off thoroughly in two or three waters, and dry with a soft linen cloth, or if frosted or chased work, with fine, clean sawdust. This process of cleaning is very effective and satisfactory, if properly employed; but any neglect or carelessness in washing off and rinsing the article, after it is removed from the solution, may result in its permanent injury, as the strong alkali, if unremoved, will, when dry, corrode or eat into, and thus irretrievably mar, the surface.

Tarnished jewelry may be speedily restored to its original color by similar process; and a little of this solution mixed with Spanish whiting will effectually remove stains or discolorations from almost any metallic surface to which it is applied.

In the arts, cyanuret potassa ranks among the most valuable chemicals known; but, being a deadly poison, too great care can not be observed in its use, as even the smallest quantity, accidentally taken into the stomach, might prove fatal. In its use it is also desirable to avoid as much as possible inhaling the odor arising from it, which to persons with diseased or weak lungs is also found to be injurious.

Section 11.—How to give to Silver a Bright Gold Tinge.

This may be done by steeping it for a suitable length of time in a weak dilution of sulphuric acid and water, strongly impregnated with iron rust. The rust impregnation may be imparted by immersing nails or other pieces of iron in the diluted acid.

Section 12.—To remove instantaneously Blueing or Coloring from Steel.

Take equal parts of muriatic acid and elixir vitriol; mix and apply with a bit of peg-wood, or whatever is most convenient, and wipe dry.

Section 13.—To neutralize the effects of Acid, when the hands or clothes become accidentally exposed to it, and to restore the Color to garments where it is partially destroyed by Acid.

Dampen the part or parts immediately, or as soon after the accident as possible, with *spirits ammonia.* This, if the fabric is not already actually destroyed, will almost instantaneously produce the desired result. This fact, though known to all chemists, may not be known to many into whose hands this work may fall, and who have occasion to use acids; hence its insertion in this place.

Section 14.—To make Lacquers for Brass and Copper.

RECIPE—2 oz. alcohol; ¼ oz. shell-lac.

Dissolve and strain through filtering paper, and while filtering keep covered to prevent evaporation. It may now be used as a transparent lacquer or varnish to protect polished brass, copper, etc., from tarnish, or may be colored at pleasure, as follows: for *yellow* use *gamboge;* for *orange* to *crimson,* use *dragon's-blood;* and for *bright red,* for watch hands, etc., use *red sanders;* or any other desired color (soluble in alcohol) may be used. First dissolve the colors in alcohol, and, when settled, mix sufficient with the lacquer to produce the color desired.

DIRECTIONS FOR USE.—Clean the article from all stains and dirt, and in cold weather warm it slightly; then apply the lacquer, quickly and evenly, with a camel's-hair brush, taking care not to go over any part a second time until the first is fully dry. The number of coatings will, to some extent, determine both the color and shade. To remove lacquer, wash in alcohol.

Section 15.—To make Red Watch Hands.

FIRST RECIPE—1 oz. carmine; 1 oz. muriate of silver; ½ oz. tinner's japan.

Mix together in an earthen vessel, and hold over a spirit lamp until formed into a paste. Apply this to the watch hand, and then lay it on a copper plate, face side up, and heat the plate sufficiently to produce the color desired.

SECOND RECIPE—25 gr. gum sandric; 10 gr. red sanders; ½ oz. *best* sulphuric ether.

Dissolve and shake well together. To be applied by dipping the hand quickly two or three times into the solution, allowing it each time to dry thoroughly. This process gives to composition and gold hands a beautiful *red,* and to steel or iron a *bright purple* color; the only drawback to it being that, if not skillfully applied, the color is liable to run into streaks.

THIRD RECIPE—(See sec. 7), "Lacquers for Brass and Copper." This lacquer, colored with red sanders, gives a very fine red color, and may be applied with ease, and a certainty of success.

Section 16.—To make Liquid Bronzes for Plaster Figures, etc.

RECIPE—5 oz. palm soap; $1\frac{1}{2}$ oz. sulphate of copper; $1\frac{1}{2}$ oz. sulphate of iron; linseed oil, or oil of turpentine, as much as required.

Dissolve the soap in rain-water in one vessel, and the sulphates of iron and copper in another vessel, and when dissolved mix the solution of sulphates with the soap solution until no further precipitate occurs; then dry the precipitate, and mix it with the linseed oil, or oil of turpentine, which ever is preferred. The sulphate of copper alone with the palm soap gives a bright green, and the sulphate of iron a yellow color. If preferred, these preparations may be made separately, and then mixed to taste.

Section 17.—How to make Polishing Powders and to free them from Gritty Matter.

Freeing from gritty matter may be accomplished by simply washing the powder in clean *rain-water* in the following manner:

Mix the powder with a plentiful supply of clean rain-water; stir it up well, and let it subside barely long enough to allow the gritty matter to settle to the bottom, and then pour off the top carefully into another vessel. Let this settle fully; then drain and evaporate to dryness. When dry, reduce to a fine powder, and put into a close-covered vessel for use. A second or third washing before drying may be resorted to if an extra fine powder is desired.

Spanish whiting and ground chalk both make excellent polishing powders when prepared in this way; to either of which the addition of a little jeweler's rouge is a decided improvement. The rouge may be mixed with the whiting or chalk at the outset, and all washed together. A very excellent polishing powder for brass and copper, to be used dry, may be made by simply washing finely ground brick-dust as above. This constitutes the powder sold very extensively not long since, under slight disguise, as *"Magic Polishing Powder."*

What is known as *French Plate Powder* is made by mixing one part of fine jeweler's rouge with twelve parts of carbonate of magnesia.

PART SECOND.

DEVOTED SPECIALLY TO PROCESSES, EMPLOYED IN WATCH REPAIRING.

Section I.—List of Trains of Watches.

SHOWING THE NUMBER OF TEETH IN THE WHEELS, LEAVES IN THE PINIONS, BEATS IN A MINUTE, AND TIME THE FOURTH WHEEL REVOLVES IN.

Trains, for Seven Teeth in the Escapement Wheel.

No. of Teeth in the Center Wheel	Teeth in 3d Wheel	Leaves in 3d Wheel Pinion	Teeth in 4th Wheel	Leaves in 4th Wheel Pinion	Teeth in the Escapement Wheel	Leaves in the Escapement Wheel Pinion	No. of Beats in one minute	No. of Seconds the 4th Wheel revolves in
72	66	6	58	6	7	6	298—	27
66	64	6	64	6	7	6	292+	31
66	64	6	63	6	7	6	287+	31
66	63·	6	63	6	7	6	283—	31
66	63	6	62	6	7	6	278+	31
66	63	6	61	6	7	6	274—	31
66	63	6	60	6	7	6	269+	31

Trains, for Nine Teeth in the Escapement Wheel.

No. of Teeth in the Center Wheel	Teeth in 3d Wheel	Leaves in 3d Wheel Pinion	Teeth in 4th Wheel	Leaves in 4th Wheel Pinion	Teeth in the Escapement Wheel	Leaves in the Escapement Wheel Pinion	No. of Beats in one minute	No. of Seconds the 4th Wheel revolves in
63	60	6	57	6	9		299+	34
66	60	6	54	6	9		297	33
63	60	6	56	6	9		294	34
66	60	6	53	6	9		291+	33
63	60	6	55	6	9		289—	34
66	60	6	52	6	9		286	33
63	60	6	54	6	9		283+	34
66	60	6	51	6	9		280+	33
63	60	6	53	6	9		278+	34
66	60	6	50	6	9		275	33
63	60	6	52	6	9	6	273	34

Trains, for Eleven Teeth in the Escapement Wheel.

No. of Teeth in the Center Wheel	Teeth in 3d Wheel	Leaves in 3d Wheel Pinion	Teeth in 4th Wheel	Leaves in 4th Wheel Pinion	Teeth in the Escapement Wheel	Leaves in the Escapement Wheel Pinion	No. of Beats in one minute	No. of Seconds the 4th Wheel revolves in
60	60	6	49	6	11	6	300—	36
60	54	6	54	6	11	6	297	40
60	56	6	52	6	11	6	230—	30
64	52	6	52	6	11	6	294—	30
58	56	6	53	6	11	6	292+	40

No. of Teeth in the Center Wheel......	Teeth in 3d Wheel.....	Leaves in 3d Wheel Pinion......	Teeth in 4th Wheel....	Leaves in 4th Wheel Pinion......	Teeth in the Escapement Wheel......	Leaves in the Escape Wheel Pinion......	No. of Beats in one Minute......	No. of Seconds the 4th Wheel revolves in...
60	54	6	53	6	11	6	291+	40
62	54	6	51	6	11	6	290—	39
58	54	6	54	6	11	6	287+	41
58	55	6	53	6	11	6	287	41
59	54	6	53	6	11	6	286+	41
60	54	6	52	6	11	6	286	40
60	55	6	51	6	11	6	286—	39
61	55	6	50	6	11	6	285—	39
63	55	6	48	6	11	6	282+	38
59	54	6	52	6	11	6	281+	41
60	54	6	51	6	11	6	281+	40
61	54	6	50	6	11	6	280—	39
56	54	6	54	6	11	6	277+	43
60	60	6	48	6	11	6	293+	36
62	54	6	52	6	11	6	295+	39
63	54	6	50	6	11	6	289—	38
63	48	6	56	6	11	6	287+	43
70	70	7	56	7	11	7	293+	36
70	70	7	48	7	11	6	293+	36
70	60	7	48	6	11	6	293+	36
60	70	6	48	7	11	6	293+	36
63	50	6	56	7	11	6	287+	40
63	63	6	50	7	11	6	289—	38
80	80	8	64	8	11	8	293+	36
80	80	8	56	8	11	7	293+	36
80	80	8	48	8	11	6	293+	36
80	70	8	56	7	11	7	293+	36
80	70	8	48	7	11	6	293+	36
80	60	8	48	6	11	6	293+	36
70	80	7	56	8	11	7	293+	36
70	80	7	48	8	11	6	293+	36
60	80	6	48	8	11	6	293+	36
84	72	8	50	8	11	6	289—	38
84	63	8	50	7	11	6	289—	38
84	54	8	50	6	11	6	289—	38
63	72	6	50	8	11	6	289—	38
63	63	6	50	7	11	6	289—	38
84	64	8	56	8	11	6	287+	40
84	56	8	56	7	11	6	287+	40
84	48	8	56	6	11	6	287+	40
63	64	6	56	8	11	6	287+	40
63	56	6	56	7	11	6	287+	40

Trains, for Thirteen Teeth in the Escapement Wheel.

No. of Teeth in the Center Wheel	Teeth in 3d Wheel	Leaves in 3d Wheel Pinion	Teeth in 4th Wheel	Leaves in 4th Wheel Pinion	Teeth in the Escapement Wheel	Leaves in the Escapement Wheel Pinion	No. of Beats in one Minute	No. of Seconds the 4th Wheel revolves in
54	53	6	52	6	13	6	298+	45
56	53	6	50	6	13	6	298—	44
59	51	6	49	6	13	6	296—	43
60	51	6	48	6	13	6	294+	42
54	53	6	51	6	13	6	293—	45
56	53	6	49	6	13	6	292—	44
56	54	6	48	6	13	6	291+	44
57	53	6	48	6	13	6	291—	43
54	52	6	51	6	13	6	287+	46
54	43	6	50	6	13	6	287+	45
50	51	6	50	6	13	6	286+	45
54	52	6	50	6	13	6	282—	46
56	51	6	49	6	13	6	281—	45
57	51	6	48	6	13	6	280—	44
52	52	6	51	6	13	6	277—	48
53	52	6	50	6	13	6	276+	46
52	52	6	52	6	13	6	293—	46
55	51	6	51	6	13	6	287	46
56	50	6	51	6	13	6	286+	46
56	52	6	48	6	13	6	280+	44
56	52	6	50	6	13	6	292+	44
60	48	6	48	6	13	6	277+	45
60	50	6	48	6	13	6	289—	43
60	54	6	60	8	13	6	292+	53
60	58	7	56	7	13	6	287+	51
60	60	8	54	6	13	6	300	44
62	56	7	56	7	13	6	296+	47
63	52	7	51	6	13	6	285	60
63	60	7	60	7	13	6	290	60
64	60	7	60	7	13	6	285	60
72	70	8	68	8	13	6	280	60
74	68	8	68	8	13	6	286+	60

Trains, for Fifteen Teeth in the Escapement Wheel.

No. of Teeth in the Center Wheel	Teeth in 3d Wheel	Leaves in 3d Wheel Pinion	Teeth in 4th Wheel	Leaves in 4th Wheel Pinion	Teeth in the Escapement Wheel	Leaves in the Escapement Wheel Pinion	No. of Beats in one Minute	No. of Seconds the 4th Wheel revolves in
54	50	6	48	6	15	6	286	48
58	48	6	46	6	15	6	290	50
48	45	6	59	6	15	6	291—	60
48	45	6	58	6	15	6	300	62
48	45	6	57	6	15	6	288	62
48	45	6	56	6	15	6	288	50

No. of Teeth in the Center Wheel. ...	Teeth in 3d Wheel....	Leaves in 3d Wheel Pinion.........	Teeth in 4th Wheel....	Leaves in 4th Wheel Pinion.........	Teeth in the Escapement Wheel.......	Leaves in the Escapement Wheel Pinion.	No. of Beats in one Minute........	No. of Seconds the 4th Wheel revolves in...
56	48	6	46	6	15	6	289—	50
63	56	7	56	7	15	7	288	50
60	56	8	58	7	15	6	288	50
62	60	8	60	8	15	6	288	50
72	64	8	50	8	15	6	288	50
72	64	8	56	8	15	7	288	50
72	64	8	64	8	15	8	288	50
52	50	6	48	6	15	6	288	50
54	48	6	48	6	15	6	288	50
72	64	8	48	8	16	6	288	50
72	80	8	64	10	15	8	288	50
72	80	8	56	10	15	7	288	50
72	80	8	48	10	15	6	288	50
63	80	7	64	10	15	8	288	50
63	80	7	56	10	15	7	288	50
63	80	7	48	10	15	6	288	50

Trains, for Seventeen Teeth in the Escapement Wheel.

64	80	8	48	10	17	6	299+	53
54	48	6	44	6	17	6	299+	50
51	48	6	45	6	17	6	295+	53
54	48	6	43	6	17	6	292+	50
48	48	6	48	6	17	6	290+	53
51	48	6	45	6	17	6	289	53
54	48	6	42	6	17	6	286—	53
48	48	6	47	6	17	6	284+	53
51	48	6	44	6	17	6	283—	53
48	48	6	46	6	17	6	278	53
48	48	6	45	6	17	6	272	53
64	64	8	64	8	17	8	290+	50
72	64	8	56	8	17	8	286—	50
64	64	8	60	8	17	8	289—	53
56	56	7	56	7	17	7	290+	53
63	56	7	49	7	17	7	286—	50
64	56	8	48	7	17	6	290+	53
80	80	10	64	10	17	8	290+	53
80	64	10	64	8	17	8	290+	53
80	64	10	56	8	17	7	290+	53
80	64	10	48	8	17	6	290+	53
80	56	10	56	7	17	7	290+	53
80	56	10	48	7	17	6	290+	53
64	80	8	64	10	17	8	290+	53
64	80	8	56	10	17	7	290+	53

Trains, for Third Wheel and Patent Seconds.

No. of Teeth in the Center Wheel	Teeth in 3d Wheel	Leaves in 3d Wheel Pinion	Teeth in 4th Wheel	Leaves in 4th Wheel Pinion	Teeth in the Escapement Wheel	Leaves in the Escapement Wheel Pinion	No. of Beats in one Minute	No. of Seconds the 4th Wheel revolves in
60	72	6	60	12	...	6	300	60
60	60	6	60	10	...	6	300	60
60	48	6	60	8	...	6	300	60
48	60	6	60	8	...	6	300	60
60	72	6	54	12	...	6	270	60
60	60	6	54	10	...	6	270	60
48	60	6	54	8	...	6	270	60
60	72	6	48	12	...	6	240	60
60	60	6	48	10	...	6	240	60
48	60	6	48	8	...	6	240	60

Trains, for Fourth Wheel Seconds, with Eleven Teeth in the Escapement Wheel.

No. of Teeth in the Center Wheel	Teeth in 3d Wheel	Leaves in 3d Wheel Pinion	Teeth in 4th Wheel	Leaves in 4th Wheel Pinion	Teeth in the Escapement Wheel	Leaves in the Escapement Wheel Pinion	No. of Beats in one Minute	No. of Seconds the 4th Wheel revolves in
48	45	6	71	6	11	6	260+	60
48	45	6	74	6	11	6	271+	60
48	45	6	76	6	11	6	279—	60
48	45	6	78	6	11	6	286	60
60	49	7	74	7	11	6	271+	60
60	49	7	76	7	11	6	279—	60
60	49	7	78	7	11	6	286	60
45	56	6	74	7	11	6	271+	60
45	56	6	76	7	11	6	279—	60
45	56	6	78	7	11	6	286	60
64	60	8	74	8	11	6	271+	60
64	60	8	76	8	11	5	279—	60
64	60	8	78	8	11	6	286	60
60	56	8	74	7	11	6	271+	60
60	56	8	76	7	11	6	279—	60
60	56	8	78	7	11	6	286	60
60	48	8	74	6	11	6	271+	60
48	48	8	78	6	11	6	286	60
48	60	6	74	8	11	6	271+	60
48	60	6	78	8	11	6	286	60
56	60	7	74	8	11	6	271+	60

Trains, for Fourth Wheel Seconds, with Thirteen Teeth in the Escapement Wheel.

No. of Teeth in the Center Wheel...	Teeth in 3d Wheel...	Leaves in 3d Wheel Pinion...	Teeth in 4th Wheel...	Leaves in 4th Wheel Pinion...	Teeth in the Escapement Wheel...	Leaves in the Escapement Wheel Pinion.	No. of Beats in one Minute...	No. of Seconds the 4th Wheel revolves in...
64	60	8	66	8	13	6	286	60
64	60	8	67	8	13	6	290+	60
64	60	8	68	8	13	6	295—	60
64	60	8	69	8	13	6	299	60
60	49	7	77	7	13	7	286	60
60	49	7	66	7	12	6	286	60
60	49	7	67	7	13	6	290+	60
48	45	6	66	6	13	6	286	60
48	45	6	67	6	13	6	290+	60
48	45	6	68	6	13	6	264—	60
48	45	6	69	6	13	6	299	60
60	56	8	66	7	13	6	286	60
80	60	10	66	8	13	6	286	60
64	75	8	66	10	13	6	286	60
48	60	6	66	8	13	6	286	60
48	75	6	66	10	13	6	286	60
45	56	6	66	7	13	6	286	60
56	75	7	68	10	13	6	295—	60

Trains, for Fourth Wheel Seconds, with Fifteen Teeth in Escapement Wheel.

No. of Teeth in the Center Wheel...	Teeth in 3d Wheel...	Leaves in 3d Wheel Pinion...	Teeth in 4th Wheel...	Leaves in 4th Wheel Pinion...	Teeth in the Escapement Wheel...	Leaves in the Escapement Wheel Pinion.	No. of Beats in one Minute...	No. of Seconds the 4th Wheel revolves in...
64	60	8	70	8	15	7	300	60
64	60	8	60	8	15	6	300	60
64	45	8	60	6	15	6	300	60
60	56	8	60	7	15	6	300	60
48	60	6	60	8	15	6	300	60
60	70	7	70	7	15	7	300	60
60	49	7	60	7	15	6	300	60
48	49	6	60	6	15	6	300	60
80	45	10	70	8	15	7	300	60
75	60	10	60	8	15	6	300	60
64	64	8	70	10	15	7	300	60
64	75	8	60	10	15	6	300	60
56	75	7	70	10	15	7	300	60
56	75	7	60	10	15	6	300	60
64	75	8	54	8	15	6	270	60
60	60	8	54	7	15	6	270	60
64	56	8	54	6	15	6	270	60
48	45	6	54	8	15	6	270	60
60	60	7	63	7	15	7	270	60

No. of Teeth in the Center Wheel	Teeth in 3d Wheel	Leaves in 3d Wheel Pinion	Teeth in 4th Wheel	Leaves in 4th Wheel Pinion	Teeth in the Escapement Wheel	Leaves in the Escapement Wheel Pinion	No. of Beats in one Minute	No. of Seconds the 4th Wheel revolves in
60	49	7	54	7	15	6	270	60
48	49	6	54	6	15	6	270	60
64	45	8	48	8	15	6	240	60
60	60	8	48	7	15	6	240	60
48	50	6	48	8	15	6	240	60
64	60	8	48	6	15	6	240	60
60	45	7	56	7	15	7	240	69
60	49	7	48	7	15	6	240	60
48	45	6	48	6	15	6	240	60
60	56	8	48	7	15	6	240	60

Trains, for Fourth Wheel Seconds, with Seventeen Teeth in Escapement Wheel.

64	60	8	51	8	17	6	289	60
64	60	8	50	8	17	6	283+	50
60	56	8	51	7	17	6	289	60
80	60	10	50	8	17	6	283+	60
75	64	10	50	8	17	6	283+	60
75	56	10	68	7	17	8	289	60 ·
75	68	10	68	8	17	8	289	60
80	75	10	68	10	17	8	289	60

Train of the American Watch Company's Watch.

64	60	8	64	8	15	7	300	60

NOTE.—By use of the foregoing set of Trains, and the rule for sizes of pinions, in sec. 2, all difficulty of calculating is obviated; and at one view, in case of the accidental loss of a wheel and pinion, may be known the correct size and count of the pinion, and number of teeth in the wheel lost.

Section 2.—Rule for determining the correct Diameter of a Pinion by Measuring Teeth of the Wheel that matches into it.

The term FULL, as used below, indicates full measure, from outside to outside of the teeth named; and the term CENTER, the measure from center of one tooth to center of the other tooth named, inclusive:

For diameter of a pinion of 15 leaves measure, with calipers, a shade less than 6 teeth of the wheel, *full.*

For diameter of a pinion of 14 leaves measure, with calipers, a shade less than 6 teeth of the wheel, *center.*

For diameter of a pinion of 12 leaves measure, with calipers, 5 teeth of the wheel, *center.*

For diameter of a pinion of 10 leaves measure, with calipers, 4 teeth of the wheel, *full.*

For diameter of a pinion of 9 leaves measure, with calipers, a little less than 4 teteh of the wheel, *full.*

For diameter of a pinion of 8 leaves measure, with calipers, a little less than 4 teeth of the wheel, *center.*

For diameter of a pinion of 7 leaves measure, with calipers, a little less than 3 teeth of the wheel, *full.*

For diameter of a pinion of 6 leaves measure, with calipers, a little over 3 teeth of the wheel, *center.*

For diameter of a pinion of 5 leaves measure, with calipers, 3 teeth of the wheel, *center.*

For diameter of a pinion of 4 leaves measure, with calipers, one half of one space over 2 teeth of the wheel, *full.*

As a general rule, pinions that lead, as in the hour wheel, should be somewhat larger than those that drive; and pinions of clocks should generally be somewhat larger proportionately than those of watches.

Section 3.—How to make Diamond Mills, Broaches, and Files, for Grinding Jewels, Enlarging Jewel Holes, Polishing Jeweled Pallets, etc.

For these purposes diamond dust is required. It may be bought in New York, Boston, Philadelphia, and occasionally in other places, ready prepared; or any one may prepare it for himself by crushing to dust a few small pieces of common or cheap diamond. This, with ordinary care, may be done, with very little or no loss of particles, in the following manner: place the diamond on a block of hard polished steel, in a suitable vessel, and cover it with water to prevent flying or scattering; then place a flat steel punch on each piece separately, and strike the punch with a mallet or hammer, with sufficient force to crush the diamond. When reduced sufficiently fine in this way, the dust may be carefully collected and dried for use; after drying, it may be graduated for different purposes by mixing it with a little watch oil; when agitated, the finest particles will float near the surface, while the coarsest pieces will sink at once to the bottom; and thus, by decanting the oil in which the dust floats, as many grades of fineness as desired may be readily obtained.

If desired, the dust may now be separated from the oil by pouring each grade separately on to a piece of clean, smooth paper; the paper will absorb the oil, or allow it to filter through, while the diamond will remain on the surface. But as this process necessarily involves more or less waste of the dust, it is generally better to leave it in the oil, and use it directly therefrom, as occasion requires; or the oil may be washed out of the dust with alcohol.

For a Diamond Mill.—Make a brass chuck or wheel, suitable for use on a foot-lathe, with a flat, even surface or face of about one and a half or two inches in diameter; then place a number of the coarsest pieces of your diamond dust on different parts of its face, and with a smooth-faced steel hammer drive the pieces of dust all evenly into the brass to nearly or quite level with the surface. Your mill thus prepared may now be used for making pallet jewels, or for grinding stone or glass of whatever kind desired. For polishing, use a bone or boxwood chuck or wheel, of similar form to your mill, and coat it lightly with the finest grade of your diamond dust and oil; with this a beautiful polish may be given to even the very hardest stone.

For Diamond Broaches.—Prepare a number of brass pins, of the size and form you desire; then stir the point of each separately into a medium grade of your diamond dust and oil until it becomes evenly covered; then place it on a smooth steel stake or anvil, and with a light hammer go carefully over the entire surface, driving the dust into the brass. A few minutes of light, even hammering in this way, taking care not to strike so hard as to dent or flatten the pin, will completely charge it with the diamond dust, which will be found to adhere with very great tenacity. Or, if preferred, the dust may be pressed into the pin by rolling it between two flat pieces of steel, instead of hammering as above. If dry diamond dust is used, first dampen the pin with oil, then dip or stir it into the dust until a sufficient quantity has adhered, and hammer or roll it as previously directed.

In using the broaches press but lightly into the jewel hole, and turn the broach rapidly with the fingers. For polishing, use a bone or ivory point, lightly coated with the finest diamond dust and oil, and while using it with the one hand, accompany the motion with a slight oscillating motion of the other hand in which the jewel is held. This will insure a more even polish to the hole, with less liability to press the jewel out of its place in the plate, · than if held firm and steady.

For Diamond File.—Charge the flat side of a piece of brass or soft iron, of suitable size and shape, with diamond dust, the same as in the case of broaches.

Section 4.—Very Simple and Valuable Process for making Pivot Files and Burnishers, of any desired Shape.

By this means the finest conical pivots may be made with the same facility us ordinary square pivots, and a more perfect finish be given to square pivots than can possibly be given by any other process.

For this purpose a smooth pivot-burnisher or a worn-out pivot-file may be used. First grind off the original teeth, and make the face perfectly level; then round slightly or bevel one of the corners the entire length of the file, so that when finished it will make the exact form of pivot you desire. When thus prepared, draw the face of the steel three or four times, at right angles, across a piece of No. 1 emery paper—which for the purpose has been previously glued on to a level block of wood—extending the cut of the emery each time over the beveled or rounded corner of the steel. A few draws across the emery in this way will cut into the steel an immense number of little grooves, and will leave on its surface a corresponding number of very fine sharp teeth, which, though shallow, will be found to act upon a pivot with surprising rapidity; and as these teeth extend alike over the rounded corner of the file, or the part intended to form the shoulder of the pivot, it follows that in its use all parts of the pivot will be equally acted upon by it at the same time. If preferred, a smooth copper plate, with dry No. 1 emery powder sprinkled over it, may be used instead of the emery paper, as above.

The reverse side of the file, for burnishing, should be made, in form, an exact counterpart of the file, and should be prepared with very fine emery paper, or with fine flour of emery, on a copper plate, instead of the No. 1 powder used in making the file. For burnishing square pivots, make the corner of your burnisher exactly square, and then finish both its face and edge on fine emery paper, or on flour of emery. When properly prepared, a burnisher thus made will finish the pivot and shoulder both perfectly down into the very corner.

In making conical pivots, first turn down with a graver, in the usual way, to about half the size you desire to make your pivot; then place it in a collar lathe, with about one half the length of the pivot resting on the collar, and apply your conical file. While resting on the collar, any desired amount of pressure may be put on the file without the slightest danger of breaking the pivot.

Ordinary flat burnishing files may be made, and those that have become damaged by use may be renewed by the foregoing process, and again made as perfect as when they came originally from the factory.

Section 5.—Greatly Improved Modes of Tempering Case and other Springs of Watches.

First Mode.—Draw the temper from the spring, and fit it properly in its place in the watch; then take it out and temper it hard in rain-water (the addition of a little table-salt to the water will be an improvement); after which place it in a small sheet-iron ladle or cup, and barely cover it with linseed oil; then hold the ladle over a lighted lamp until the oil ignites. Let it burn until the oil is nearly, not quite, all consumed; then re-cover with oil and burn down as before, and so a third time, at the end of which plunge it again into water. Main and hair-springs may, in like manner, be tempered by the same process. First draw the temper, and properly coil and clamp to keep in position, and then proceed the same as with case-springs.

Second Mode.—Draw the temper, fit the spring into its place, and temper hard the same as in first process; then polish the small end so as to show the change of color when heated; lay it on a piece of copper plate, and hold the plate over a lighted lamp, so that the flame may strike directly under the heaviest part of the spring. When the plate becomes sufficiently heated to blue the steel, remove it from the heat and let it cool gradually, the spring still remaining on the plate; then again remove the color from the end of the spring, and re-blue it in like manner; and so a third time. While cooling, care should be taken not to allow a draft of cold air to blow upon it, and thus cool it too rapidly.

Both these processes make springs of the very best quality, and that can be relied upon with the utmost confidence.

Section 6.—How to Blue all the Screws of a Watch together, evenly.

Make a small brass or copper cap or barrel, similar to a main-spring barrel, but somewhat larger in diameter, and of barely sufficient depth to allow the screws to rest upon the head or cover, points downward, without touching the bottom. Then solder on to one side a handle five or six inches long, by which to hold it, and fit the cover so suugly to it that it will touch evenly all around, or otherwise solder it around the edge carefully with brass or silver solder. Then drill through the cover a number of holes corresponding to the number of screws you desire to blue; and, placing a screw in each hole, with the point downward, and the head resting on the plate or cover, hold the cup over the flame of your lamp, with the flame striking as near the center of the bottom as may be, until the cup becomes sufficiently heated to blue the screws to the shade desired.

Section 7.—How to draw the Temper from Cylinders, Staffs, Pinions, etc., and to retemper them perfectly without changing the Color, or causing them in any degree to Warp or Spring.

Make a hollow iron or steel cylinder, of about the diameter of a common clock key, leaving the one end closed; then put the article that you desire to temper, or from which the temper is to be drawn, into the cylinder, together with sufficient brass filings to fill up the entire cavity; then plug up or stop the open end so as perfectly to exclude the external air. This done, if the temper is to be drawn merely, heat the cylinder, and let it cool gradually before removing the plug; or if it is to be tempered, heat to the desired degree, and then plunge the cylinder with its contents into water. The temper may now be graduated at pleasure by removing the color from a part of the cylinder, and then heating it again until blued; and so any desired number of times. On taking out the plug and removing the article from the cylinder, it will be found, in appearance, in precisely the same condition as when it was put in.

The principle consists simply in excluding the external air from the article while undergoing the changes from heat to cold.

Section 8.—To draw the Temper from any desired part of small Steel articles, without affecting the other parts.

Drill a hole through a narrow strip of brass, and slip it on the article, or grasp the part where you desire to draw the temper with a pair of common tweezers, and then direct the flame, with your blow-pipe, on to the brass or the tweezers, instead of the article itself; or, in case of a flat surface, press the end of a brass wire, of suitable size, against the spot where the temper is to be drawn, and then heat the wire as above.

When it is desired to protect perfectly the parts immediately surrounding that where the temper is to be drawn, coat them thickly with dampened plaster of paris; then proceed in other respects as before directed. In all ordinary cases, however, this latter precaution will be found wholly unnecessary.

Section 9.—To Make and Temper Drills, Gravers, etc.

Always use the best quality of steel, and, heating each time to a cherry red, hammer it well two or three times; with the view, in addition to giving it proper form, to compact the steel, and to lay the grain, nearly as possible, in some particular direction. In making drills especially, after once flattening, never turn and hammer it on the edge, otherwise the grain of the steel becomes crossed and rendered liable to crumble when hardened.

For tempering, use resin, beeswax, or quicksilver; or in some cases a solution of cyanuret potassa and rain-water may answer a better purpose than any of the other articles named.

For sharpening, first use oil-stone, and then finish off with Scotch graystone, which will be found to give a smoother and finer edge than can be given by any other means.

Section 10.—How to Drill into Hard Steel without drawing the Temper.

But few instances arise in which it is desirable to attempt to drill into steel so hard that it can not be acted upon with a file; and in all cases of very hard steel, when the temper can be slightly drawn without injury to the article, it is better to do so. When this can not be done, however, the following, for drilling with a a steel drill, is probably the best process known.

Make your drill oval in form, instead of the usual pointed shape, and temper, as per sec. 9, as hard as it will bear without breaking. Then roughen the surface where you desire to drill with a little diluted muriatic acid, and instead of oil use turpentine or kerosene (the kerosene for this purpose may be improved by dissolving in it a little gum camphor) with your drill. In operating, keep the pressure on your drill firm and steady; and if the bottom of the hole should chance to become burnished so that the drill will not act, as sometimes happens, again roughen with diluted acid, as at the first; then clean out the hole carefully, and proceed as before.

A little patient perseverance in this way will overcome all difficulties, and enable you to drill into the hardest staff or pinion with entire success.

Section 11.—How to Polish Steel.

For this purpose there is a white French polishing powder, sold in New York and some other places, the ingredients of 'which have not been made public, that is perhaps better than any thing else that can be used. But as this is sometimes difficult to obtain, crocus or oxid of tin may be used in its stead. In using crocus, for the best possible result, first graduate it in oil same as in preparing diamond dust, (see sec. 3, part second,) and apply it to the steel by means of a piece of soft iron or bell metal, made proper form, and prepared with flour of emery, same as for pivot burnishers. Use the coarsest of the crocus first, and finish off with the finest.

The crocus, or whatever polishing powder is used, should always be carefully protected from dust or dirt. To iron or soft steel a better finish may be given by burnishing than can be imparted by the use of polishing powder of any kind whatever.

Section 12.—Best Mode of Case-hardening Iron.

Heat the article to a red heat; then stir the part you desire to harden in cyanid of potash (powdered), and plunge it into cold water. If not found sufficiently hard the first trial, repeat the operation, increasing the degree of heat. It will come out smooth and hard as tempered steel. If you desire to harden to any considerable depth, put the article into a crucible with cyanid of potash, cover over and heat altogether, then plunge into water. This process will harden perfectly to the depth of two or three inches.

Section 13.—How to File or Grind perfectly Level.

The most perfect way to do this is to place the article in a lathe, resting on pivots at the two extreme centers, and then file or grind square across it. (In this way it will be found impossible to file other than level.) A simple substitute for the lathe, however, may be found by poising the article on the end of the finger, in such position that it will sway readily, as the handle of the file may be elevated or depressed while operating.

Section 14.—Very simple Modes of Lengthening Levers of Anchor Escapement Watches without Hammering or Soldering.

Cut square across with a screw head file, a little back from the point above the fork, and when you have thus cut into it to a sufficient depth, bend forward the desired distance the piece thus partially detached. In the event of the piece snapping off while bending—which, however, rarely happens—file down the point level with the fork, and then drill a hole at base of the fork, and insert a pin—English lever style.

Section 15.—To reduce the Strength of Hair-springs of Watches without detaching from Collets, and without changing the Color of the visible parts of the Spring.

First slip the collet off from the staff or cylinder, without removing it from the spring; then put a little diluted nitric or sulphuric acid into a watch-glass, and immerse therein the inner coil or coils only of the spring. This may be done by holding down the center of the spring with a bit of peg-wood; then, seizing the outer coil with a pair of tweezers, lift the spring into the form of an inverted cone, and, while in this position, immerse the center coil or coils in the acid. When sufficiently reduced in strength, wash off first in water, then in alcohol, and dry with tissue paper; then dip it into sulphuric ether, and dry by evaporation.

Or, if preferred, the following is a good process: First detach the spring from the collet; then pass a bit of round peg-wood through the center coil, and fasten it in your vice. This done, flatten a piece of soft steel wire so that it will pass between the coils of the spring, and with this and pulverized oil-stone and oil grind off the center coil, or that immediately surrounding the peg-wood. Five or six rubs thus, with the wire and pulverized oil-stone, will make a difference of ten to fifteen minutes per day in the running of the watch.

By both these processes, simple in their operation, the quality and value of the spring remain wholly unimpaired.

Section 16.—To put Teeth into Wheels of Watches or Clocks without Dovetailing or Soldering.

Drill a hole, somewhat wider than the tooth, square through the plate, a little below the base of the tooth, and, with a saw of the same thickness as the tooth, cut from the edge of the wheel square down to the hole already drilled. Then flatten a piece of wire so as to fit snugly into the cut of the saw, and with a light hammer form a head on it like the head of a pin. When thus prepared, press the wire or pin into position in the wheel, the head filling the hole drilled through the plate, and the end projecting out so as to form the tooth. Then with a sharp-pointed graver cut a small groove each side of the pin from the edge of the wheel down to the hole, and with a blow of your hammer spread the face of the pin so as to fill the grooves just cut. Repeat the same operation on the other side of the wheel, and finish off in the usual way. The tooth will be found perfectly riveted in on every side, and as strong as the original one, while in appearance it will be quite equal to the best dovetailing.

Section 17.—To Polish Wheels perfectly without Rounding Edges of the Arms or Teeth, or Clogging with Polishing Powders.

Take a flat burnishing file, warm it over a spirit lamp, and coat it lightly with beeswax. When cold, wipe off as much of the wax as can be readily removed, and with your file thus prepared polish the wheel, resting the wheel while polishing on a piece of cork. The finish produced will be quite equal to the finest buff polish, while there will be no clogging, and the edges of the arms and teeth will remain perfectly square.

Section 18.—Various Modes of preventing Chains of Watches from running off the Fusee.

The proper mode of doing this must of course depend upon the nature of the difficulty. If it results from the fusee not being upright, of course the proper remedy is to upright it. If the fault is in the chain, it may generally be obviated by shifting the chain end for end. If, however, the fault is in the fusee, the result either of breakage or wear, other means have to be resorted to. In some cases, filing off a very little from the outer lower edge of the chain, evenly the entire length, so as to make a slightly acute angle of the lower inner corner, will effectually prevent running off. The effect of filing thus is to cause the top part of the chain to draw a little off from the body part of the fusee, and the lower part from the same cause to draw close in against it. If, however, the fusee is so damaged as not to have left adequate support for the chain, the only way is to recut it, or to throw it out of perpendicular, with the top of winding arbor declining somewhat from the barrel. The effect of throwing thus out of perpendicular will be to cause the chain necessarily to draw close down against the body part of the fusee; in which case, even though the outer rim or guard be entirely gone, the chain will not run off.

Section 19.—To alter the Depth of Escapement of Lever Watches.

1. *Common Watches.*—Knock out the staff, and with a small file or with emery powder cut the hole oblong, in the opposite direction to what you desire to move the pallets; then replace the

staff, wedge it into position with a bit of brass, and carefully solder.

2. *When the staff is put in with a screw, as in Swiss watches.*— Take out the staff, separate the lever from the pallets, and countersink the pin-holes on the inner sides; then replace the parts, and, with the edge of the lever resting on a steel stake, place a punch of proper form on the edge of the pallets, and by a smart blow of your hammer bend the pins, uniting the pallets and lever, in the direction and to the extent you desire.

3. *Very fine Watches.*—Put in a new staff, and in making your pivots throw them to one side from the center, the exact distance you desire to alter the escapement; or, in other words, make what are known as eccentric pivots to your staff.

Section 20.—Rules for determining the correct length of the Lever, size of Ruby Pin Table, size of the Pallets, and Depth of Escapement of Lever Watches.

A lever, from the guard point to the pallet staff, should correspond in length with twice the diameter of the ruby pin table; and when a table is accidentally lost, the correct size thereof may be known by measuring half the length of the lever between the points above named. For correct size of pallet, the clear space between the pallets should correspond with the outside measure, on the points, of three teeth of the escapement wheel. The only rule that can be given, without the use of diagrams, for correct depth of the escapement, is to set it close as it will bear, and still free itself perfectly when in motion. This may be done by first placing the escapement in your depthing tool, and there setting it to the correct depth. Then by measuring the distance between the pivots of the lever staff and escapement wheel, as now set, and the corresponding pivot holes in the watch, you determine correctly how much the depth of the escapement requires to be altered.

Section 21.—Simple Rule for putting Cylinder Escapement and Skeleton Lever Watches in Beat.

All watchmakers know that, to be in proper beat, the hair-spring of a watch must be so placed that, in cylinder escapements, the square cut across the center of the cylinder, when at rest, will stand in line with the stud or outer fastening of the hair-spring, and at exact right angles with a line from the center of the escapement wheel to the center of the cylinder; and in lever watches, so that the ruby pin will stand in direct line between the balance staff and fulcrum point or staff of the lever. In all skeleton

watches, therefore, where the regulator is placed on the cock above
the balance, the simplest mode of securing these results is to put
the cock into position in the watch, and move the regulator thereon,
so that, if a cylinder, it will stand in line with the center of the
escapement wheel, and, if a lever, with the center of the pallet
staff; then, on turning over the cock in order to fasten the hair-
spring in its place, the position of the regulator will indicate the
direction of the escapement wheel from the cylinder, or of the lever
staff from the balance staff; and all that is necessary in order to
put the watch in perfect beat without further trial, will be to fasten
the hair-spring so as to cause the square cut of the cylinder to
stand at right angles with the regulator, as placed, or the ruby
pin of the lever in direct line therewith.   In cases where the
regulator is placed on the plate of the watch, instead of the cock
the foregoing rule will, of course, not apply, and other means
have to be employed to determine when the collet is in proper
position on the staff.

Section 22.—Simple Mode of Tightening a Cannon Pinion on the Center Arbor when too loose.

Grasp the arbor lightly with a pair of cutting nippers, and by
a single turn of the nippers around the arbor cut or raise a small
thread thereon.   This will be found to work more satisfactorily
and is less trouble than putting in a hair, as practiced by many.

Section 23.—To loosen Screws that have become Rusted in, and to remove Rust, Hardened Grease, etc., from Iron and Steel, and to protect them from Rust.

For cleaning purposes, etc., kerosene oil or benzine are probably
the best things known.   Where articles have become pitted by
rust, however, these can of course only be removed by mechanical
means, such as scouring with fine powder, or flour of emery and
oil, or with very fine emery paper.   To prevent steel from rusting,
rub it with a mixture of lime and oil, or with mercurial ointment;
either of which will be found valuable.

Section 24.—How to restore Magnetized Iron or Steel to its normal condition, and to expel Quicksilver from Metals that have become coated or impregnated therewith.

To withdraw magnetism from steel, etc., cover the article with
the juice of common garlic, and then warm it over a spirit lamp.
It need not be heated so much as to draw the temper or to blue
the steel.   Quicksilver is expelled by heating only.   A degree of

heat considerably below that necessary to redden the metal will expel the quicksilver in the form of vapor.

Section 25.—Simple mode of determining the exact Focal Distance of Spectacle Glasses.

Place the end of a measure of thirty or forty inches in length against a smooth wall, or other suitable ground, in plain view of some well-defined object a few rods distant, as for instance a building or window on the opposite side of the street. Then place the edge of your lens on the measure, and move it backward or forward until a spectrum is formed, or, in other words, until a clear and distinct outline of the distant object is produced on the ground against which your measure rests. This point will represent, sufficiently near, for all practical purposes, the exact focal distance of the lens, and will correspond in inches with the number on all properly marked convex spectacles.

For mending fine steel spectacle frames, use the best gold in preference to silver or brass solder.

PART THIRD.

EMBRACING PROCESSES THAT REQUIRE THE FUSING OR MELTING OF THE ARTICLES EMPLOYED.

Section I.—How to make Gold, Silver, Brass, Tin, Zinc Solders.

GOLD SOLDER, FOR FOURTEEN TO SIXTEEN-CARAT WORK.—RECIPE—Gold coin, 1 dwt.; pure silver, 9 gr.; pure copper, 6 gr.; brass, 3 gr.

Melt together in charcoal fire; and for finer work, as eighteen to twenty-two carat, use a larger proportion of gold coin. Or the following may be used in its stead, when a darker-colored solder is desired:

RECIPE—Gold coin, 1 dwt.; pure copper, 8 gr.; pure silver, 5 gr.; brass, 2 gr.

SILVER SOLDER.—FIRST RECIPE—Silver, two parts; brass, one part.

SECOND RECIPE—*Silver and common brass pin tongues, equal parts*. This latter solder flows at a much lower temperature than the first; and the use of pin tongues in its preparation will be found to answer a much better purpose than unprepared brass would, in the same proportion.

BRASS SOLDER.—RECIPE—Two parts brass; one part zinc.

TIN SOLDER.—RECIPE—Two parts zinc; one part lead.

The above is the ordinary formula for tin or soft solder, but for a solder that will flow at a very low temperature, not exceeding that of boiling water, use five parts bismuth; three parts tin; two parts lead.

ZINC SOLDER (*Liquid for causing tin solder to flow on steel, iron, etc.*)—RECIPE—Best muriatic acid, 2 oz.; sheet zinc, as much as the acid will dissolve; sal ammoniac, ¼ oz.; rain-water, 1 oz.

First dissolve the zinc in the acid, and then add the sal ammoniac and water.

Section 2.—To make Cement for repairing Fractured Jet and Stone Jewelry, and other similar goods.

Dissolve sufficient shell-lac in ninety-eight per cent. alcohol, to make it about the consistency of prepared glue. Apply this to the fractured edges; then press them closely together, and heat over a spirit lamp sufficiently to evaporate or expel the alcohol.

Section 3.—Chinese Cement for Fractured Glass, China, Earthenware, etc.

One of the very best cements known for the purposes indicated. It may also be used with equal success for repairing jet and stone jewelry, etc., by coloring it to the shade desired.

RECIPE—White glue, 4 oz.; pure isinglass, 1 oz.; finely ground dry white lead, ¼ oz.; clear water, ½ pt.; alcohol, 2 oz.

Heat the water, and dissolve in it the glue and isinglass, taking care not to scorch or burn them in the operation. Then remove from the fire, and, while cooling, mix in first the white lead and then the alcohol. Incorporate all thoroughly together, and bottle for use.

DIRECTIONS FOR USE.—Warm the cement until it becomes liquified, and warm slightly the article to be mended; then apply a thin coating of the cement evenly to the fractured edges, and press them closely together until they become firmly set. After mending, the article should be allowed to stand for a few days before using.

Section 4.—How to protect Stone and Paste Set-rings, etc., from Damage by Heat while mending.

Cover the head or set part of the ring, or other article, with a thick coating of dampened plaster of paris. You may then proceed with your mending without the least danger of damaging the settings; or, in all ordinary cases, simply imbedding the head or top of the ring in a piece of green apple or potato will answer the same purpose, with much less trouble. A light coating of dampened plaster of paris will, if properly applied, also protect fine Etruscan jewelry, etc., from change of color while mending; which is often a matter of very considerable importance.

Section 5.—To Refine Gold, Silver, and Copper, and to separate them from each other.

If in detached pieces, and mixed with baser metals, as in the case of scraps, filings, etc., melt all together in a crucible, with saltpeter and common potash. Use charcoal in melting, and leave in the fire about one hour. Then pour out, and when cold, swedge with a hammer and roll very thin; then cut into narrow strips, and coil slightly to prevent laying down flat. When thus prepared, immerse the coiled strips in nitric acid, same as for dissolving silver only. (See sec. 1, part first.) The acid will eat out or dissolve the silver and copper, holding them in solution, while the gold will be found in the form of a yellow powder at the bottom of the vessel. When fully dissolved, in this way, drain off the silver and copper solution into another vessel, and wash and rinse the gold sediment thoroughly with clean water, pouring the water each time of washing into the vessel containing the solution previously poured off. After washing thus, and drying, melt the gold again in a crucible, and run it through a sieve into water, (by this means it is precipitated in small lumps or nuggets, convenient for mixing with other metals,) or into bars, to suit taste or convenience.

The water used in washing the gold, and poured into the silver and copper solution, should not exceed in proportion, one quart of water to two ounces of the acid first used; as this is about the proportion of water to be added preparatory to collecting the silver in solution; which may now be done by immersing in the solution a sheet of clean copper, on which the silver will collect, but may be readily scraped off. After collecting and precipitating, in this way, all the silver in solution, drain off the copper solution, now remaining, into another vessel, and then wash and rinse the silver with clean water, and dry and melt it the same as the gold. The copper now in solution may be collected by immersing in it a piece of iron on which the copper will collect, same as the silver did on the copper plate. Thus you have the gold, silver, and copper all separate, and in a perfectly pure state.

For mixing of gold or for alloying it with silver or copper, the foregoing is the best possible form or state in which they can be had; for, being in themselves perfectly pure, you may know with certainty—what is often very desirable—the exact proportion of each that you are using.

Section 6.—To make best Eighteen-carat Gold, for Rings, etc.

FORMULA—18 parts pure gold; 4 parts pure copper; 2 parts pure silver; or, as a near approximate to the above, the following: 19½ gr. gold coin; 3 gr. pure copper; 1½ gr. pure silver.

Gold coin already containing about 2 parts in 24 of silver and copper alloy, the foregoing are probably as near the exact proportions of alloy that should be used with coin as can be arrived at.

Nugget gold varies in quality generally from twenty to twenty-three, and sometimes twenty-three and a half (or even finer) carats—generally a large proportion, and sometimes the entire of the alloy, being silver.

20 gr. *pure* gold to 4 gr. pure silver, gives a beautiful green-colored gold, for leaves of pins, etc., and to alloy with copper only gives a reddish color to the gold.

Section 7.—To make Cheap Gold.

For grades of gold below eighteen carat, the taste of the operator will generally guide him with sufficient accuracy; bearing in mind that the proportions of alloy to be used should be about in the ratio of two parts copper to one of silver. Brass should never be used in alloying gold, except for solders, as the zinc contained in it does not combine satisfactorily with the gold.

Formula, for best twelve-carat gold: 25 gr. gold coin; 13½ gr. pure copper; 7⅓ gr. pure silver.

For a very cheap red gold the following is a good formula: 18 parts copper; 4 parts gold; 2 parts silver. This makes a four-carat gold that answers a good purpose for cheap rings, pin tongues, etc.

Section 8.—Imitations of Gold.

FIRST FORMULA—Platina, 4 dwt.; pure copper, 2¼ dwt.; sheet zinc, 1 dwt.; block tin, 1¾ dwt.; pure lead, 1½ dwt.

If the composition should be found too hard or brittle for practical use, re-melting it with a little sal ammoniac will generally render it malleable as desired. For all purposes of melting, charcoal should be used, instead of stone-coal, which always renders the gold or composition brittle and unmalleable.

SECOND FORMULA—Platina, 2 parts; silver, 1 part; copper, 3 parts.

These compositions, when properly prepared, so nearly resemble pure gold, that it is extremely difficult to distinguish them therefrom. A little powdered charcoal mixed with metals while melting will be found of service.

Section 9.—Best Oreid of Gold.

FORMULA.—Pure copper, 4 oz.; sheet zinc, $1\frac{3}{4}$ oz.; magnesia, $\frac{5}{8}$ oz.; sal ammoniac, $1\frac{1}{32}$ oz.; quick lime, $\frac{9}{32}$ oz.; cream tartar, $\frac{7}{8}$ oz.

First melt the copper, at as low a temperature as it will melt, then add the zinc, and afterward the other articles, in powder, in the order named.

Section 10.—Very superior Bushing Alloy, for Pivot Holes, etc.

RECIPE—3 dwt. gold coin; 1 dwt. 20 gr. silver; 3 dwt. 20 gr. copper: 1 dwt. palladium.

This alloy, for the purpose named, is one of the very best known. It is very hard, producing but slight friction when brought into contact with steel; will not corrode; and in practical use is little inferior to the finest jeweling.

Section 11.—To make Alloyed and Imitation Silver, for Medals, etc.

FIRST FORMULA—Pure silver, 3 oz.; copper, $\frac{1}{4}$ oz.; brass, 2 oz.; bismuth, 1 oz.; saltpeter, 2 oz.; table-salt, 2 oz.; white arsenic, 1 oz.; common potash, 1 oz.

Melt together in charcoal fire, adding a little borax to make it run readily.

SECOND FORMULA—Pure copper, 1 dwt.; block tin, 25 dwt.; pure antimony, 2 dwt.; pure bismuth, $\frac{1}{4}$ dwt.

Section 12.—To make White Metal.

FORMULA—Brass, 2 oz.; lead, $2\frac{1}{2}$ oz.; block tin, $2\frac{1}{2}$ oz.; bismuth, $1\frac{1}{2}$ oz.; antimony $\frac{1}{8}$ oz. Melt all together.

Section 13.—To make best German Silver.

RECIPE—Copper, 25 parts; zinc, 15 parts; nickel, 10 parts.

Section 14.—To make Queen's Metal.

RECIPE—Block tin, $2\frac{1}{4}$ lb.; lead, $\frac{1}{4}$ lb.; antimony, $\frac{1}{4}$ lb.

PART FOURTH.

GENERAL PRICE LIST FOR WATCH REPAIRING; PRESENTING, IN TWO
SEPARATE COLUMNS, A FAIR AVERAGE OF THE PREVAILING PRICES
THAT OBTAIN IN DIFFERENT SECTIONS OF THE UNITED STATES AT
THE PRESENT TIME—THE HIGHER RATES APPLYING TO THE SOUTH,
AND TO A FEW OF THE LARGER CITIES AT THE NORTH, AND THE
LOWER RATES TO MOST OF THE OTHER SECTIONS.

ARTICLES.	AVERAGE RATES NORTH.		AVERAGE RATES SOUTH.	
A.				
Arbors, barrel, where the ratchet goes on a square, new......................	$1 50 to	$2 00	$2 00 to	$2 50
Arbors, barrel, where the ratchet goes on with screws........................	2 25 to	2 50	3 00 to	3 50
B.				
Barrels, verge and English lever, new	2 00 to	2 50	4 00 to	5 00
Barrels, Swiss and American lever and lepine, new..............................	3 00 to	3 50	6 00 to	7 00
Bushing, with brass, each hole.........	40 to	50	50 to	1 00
Bushing, with bushing alloy.............	75 to	1 00	1 00 to	1 50
Bushing, main wheel hole, with alloy	1 50 to	2 00	2 00 to	2 50
C.				
Chains, fusee, common, new............		1 00		2 00
Chains, fusee, medium, new............		1 50		2 50
Chains, fusee, best English, new........	2 50 to	3 00		3 00
Chains, fusee, mending and new hook, each		50	1 00 to	1 50
Cleaning common verge watches......		1 00		2 00
Cleaning lever and lepine watches.....		1 50	2 50 to	3 00
Cleaning chron. and duplex watches....	2 50 to	3 00	4 00 to	6 00
Cleaning rep'ting and musical watches		3 00	5 00 to	7 00
Clicks, barrel, short, new.................		75		1 00
Clicks, barrel, self-acting, new.........		1 50	2 00 to	2 50
Clicks, fusee, new.........................	40 to	50	50 to	1 50
Clicks, maintaining, new.................		1 50		2 50
Collets, balance-spring, new............	40 to	50	75 to	1 00
Cylinders, Swiss and English, new....	3 50 to	4 00	5 00 to	7 00
D.				
Dials, common, new.......................		2 00		3 00
Dials, with plain seconds or with key-hole, new.................................	2 50 to	3 00	3 00 to	4 00
Dials, English sunk seconds............	4 00 to	5 00	5 00 to	6 00
Depths, escapement, altering of........	50 to	2 00	1 00 to	3 00
Depths, wheels, altering of..............		1 00	1 50 to	2 00

ARTICLES.	AVERAGE RATES NORTH.		AVERAGE RATES SOUTH.	
F.				
Forks, lever, new......................	$3 00 to	$4 00	$3 50 to	$5 00
Followers, English potance. new.......		50		75
Followers, Swiss potance, with screws, new...		1 00	1 50 to	2 00
Fusees to common verge watches, new	2 50 to	3 00		3 00
Fusees to English lever watches, new..	3 50 to	4 00	3 50 to	5 00
Fusees, repairing, to prevent chain running off....................................	75 to	1 50	1 00 to	2 50
J.				
Jewels, common cap, new...............		75	1 00 to	2 00
Jewels, diamond cap, new..............		1 50	2 00 to	3 00
Jewels, slip, new.......................	1 00 to	1 50	1 00 to	2 00
Jewels, plate, hole, for center, third, fourth, and escapement wheels......	1 50 to	2 00	1 50 to	00
Jewels, plate, hole, for main wheel....		2 50	2 50 to	3 00
Jewels, replacing, with bushing alloy, in main wheel...........................	2 00 to	2 50	2 50 to	3 00
Jewels, pallet, each, new...............	1 50 to	2 00		2 00
Jewels, English hole, set with screws, new	2 50 to	3 00		3 00
Jewels, ruby pin, new...................		1 50		1 50
Jewels, duplex roller, new..............		4 00	4 00 to	5 00
P.				
Pallets, detached and patent lever, new	3 00 to	5 00	3 00 to	5 00
Pinions, cannon, new...................	1 50 to	2 00	3 00 to	5 00
Pinions, all others, each, new..........		2 00	4 00 to	5 00
Pivots, pallet staff, each, new..........		1 00	1 00 to	1 50
Pivots, all others, each, new............	1 25 to	1 50	1 50 to	3 00
Plugs, cylinder, each, new..............		1 50	2 00 to	3 00
R.				
Ratchets, barrel and fusee, common, each, new........	50 to	75	50 to	1 00
Ratchets, barrel, to levers and lepines, new	75 to	1 50	75 to	1 50
Ratchets, maintaining, new.............	1 50 to	2 00	1 50 to	2 00
S.				
Screws, all kinds, new...................	25 to	50	50 to	1 00
Springs, common. main, new...........		1 50		2 50
Springs, lever and lepine main, new...	1 50 to	2 00		3 00
Springs, balance or hair, common, new	75 to	1 00	1 25 to	1 50
Springs, balance or hair, best, new....	1 00 to	1 50	1 50 to	2 50

ARTICLES.	AVERAGE RATES NORTH.		AVERAGE RATES SOUTH.	
Springs, bal. or hair, isochronous, new	$2 50 to	$3 00	$3 00 to	$3 50
Springs, balance or hair, English chronometer, new......................		5 00	5 00 to	7 00
Springs, catch and lifting, each, new..	1 50 to	2 00	3 00 to	4 00
Springs, fusee click, new.................	50 to	75	75 to	1 00
Springs, click, lepine and lever, straight, new		1 00		1 00
Springs, click, lepine and lever, circular, new.......................		1 50	1 50 to	2 00
Springs, maintaining click........... ...		1 00	1 00 to	1 50
Springs, bal. or hair, altering set of...		50	50 to	75
Springs, bal. or hair, reduc'g strength of	50 to	75	75 to	1 00
Springs, main, new hook to.............	50 to	75	75 to	1 00
Staff, balance, new.......................		3 00		5 00
Staff, pallet, new.......................		2 00		3 00
Staff, rack lever and verticle, each, new		3 00		5 00
Staff, rack lever balance, new..........		5 00		7 00
Staff, duplex............................		5 00	8 00 to	9 00
Stops, lepine, each piece, new...........	50 to	75	75 to	1 00
Stops, fusee, new.......................	75 to	1 00		1 00
Studs, balance spring, new.............		50		75
Studs, cap, new..........................	50 to	75	1 00 to	1 50

T.

ARTICLES.	NORTH		SOUTH	
Teeth, putting in two or more together, each..		25		50
Teeth, putting in one only...............		50		1 00

V.

ARTICLES.	NORTH		SOUTH	
Verges, French and English, new......		2 00	2 50 to	3 00

W.

ARTICLES.	NORTH		SOUTH	
Wheels, common main, new............		2 00	2 50 to	3 50
Wheels, lever and lepine main, new...	3 00 to	4 00	4 00 to	5 00
Wheels, center, 3d and 4th, each, new	2 00 to	2 50	2 00 to	3 00
Wheels, escapement, lever.............		2 50		3 00
Wheels, escapement, duplex and cylinder, each, new........................	3 00 to	3 50	4 00 to	5 00
Wheels, hour, new.......................		1 25		2 00
Wheels, minute, without pinion, new		1 25		2 00
Wheels, minute, with pinion, new.....		2 50		3 00
Wheels, balance, plain, new	1 50 to	2 00	2 50 to	3 50
Wheels, balance, expansion, new......	4 00 to	5 00	5 00 to	7 00
Wheels, balance, expansion, adjusted to heat and cold.....................	10 00 to	12 00	12 00 to	14 00

CPSIA information can be obtained
at www.ICGtesting.com
Printed in the USA
LVOW13s1709090517
533870LV00023B/506/P